PREPARE FOR LIFTOFF

THE BOOK TO GET YOU WHERE YOU'RE GOING

NOELLE HOFFMAN

It is my great pleasure to inform
you that you have been selected
to participate in astronaut training.
Your proven abilities assure me that
you will be the perfect candidate for
this upcoming mission.

You hold in your hands an important
manual that will prepare you for
liftoff. Don't be afraid to markup the
pages with notes, drawings, dirt or
anything for that matter.

Good luck in your future endeavors.

PREPARE FOR LIFTOFF

ASTRONAUT PROFILE

NAME		AGE	
HEIGHT		WEIGHT	

IMAGE

CONTACT

FIGURE OUT WHERE YOU'RE GOING...

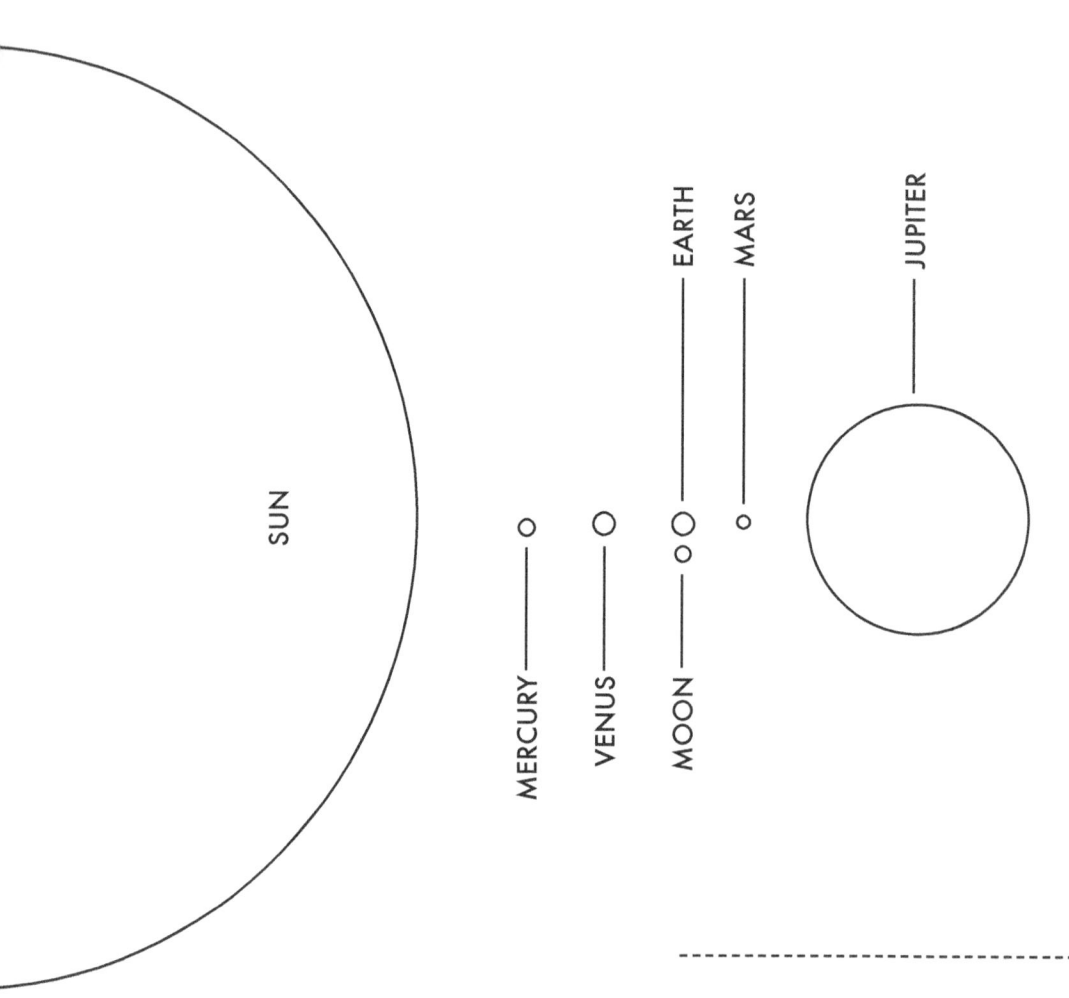

AND MAP OUT YOUR JOURNEY!

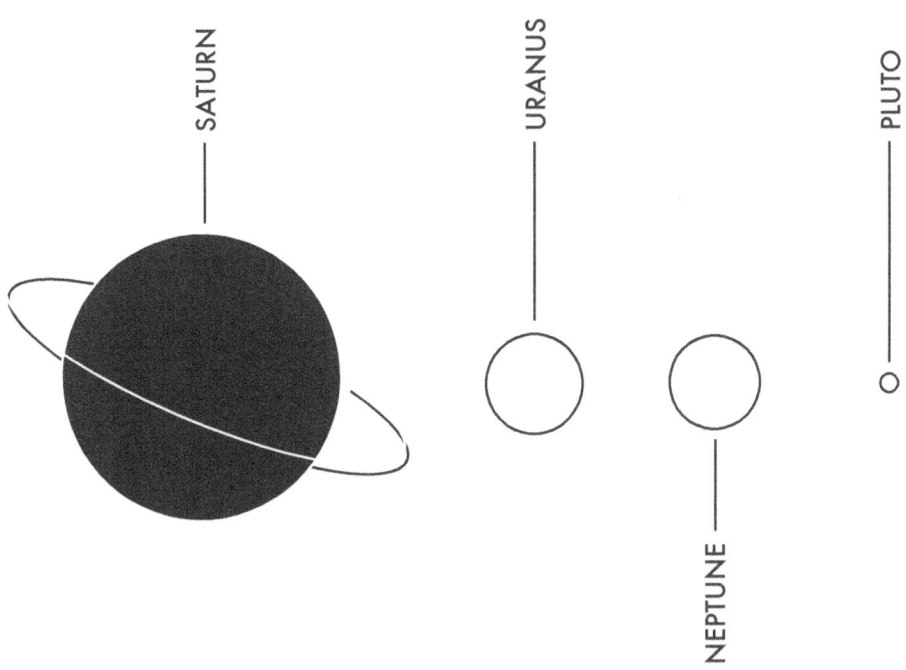

SATURN

URANUS

PLUTO

NEPTUNE

The furthest planet from Earth, Pluto, is 3.6 billion miles away.

BUT WHAT WILL YOU PACK?

Remember to pack light, one pound over
weight and your mission could be ruined.

PACKING LIST

☑ Prepare for Liftoff Training Manual

☐

☐

☐

☐

☐

☐

☐

☐

FILL THIS PAGE WITH STARS

Our galaxy alone contains over
100 billion stars! That's a lot.

YOU'LL NEED THIS PAGE TOO

A DAY IN ZERO GRAVITY

Write about zero gravity and how it
could turn your day upside down.

_____

_____

_____

_____

_____

_____

IT'S GETTING AWAY!

Draw all of your stuff floating
around in zero gravity.

HOW WILL YOU GET TO SPACE?

While in orbit, a space craft travels around Earth
at a speed of about 17,500 miles per hour.

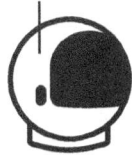

DESIGN YOUR SPACE SUIT

A space suit weighs about 280 pounds.
That's heavier than most refrigerators!

NIGHT SKY OBSERVATIONS

The night sky is full of amazing sights.
What can you see?

TIME DATE	SKY LOG
	SHOOTING STAR
	THE MOON
	JUPITER
	BIG DIPPER
	NORTH STAR
	ORION'S BELT
	THE SPACE STATION
	AN AIRPLANE
	A COMET'S TAIL
	A METEOR SHOWER

STOP AND DO SOME CRUNCHES

It's important to be in good physical shape
so you don't pass out when you launch.

I DID _____ CRUNCHES

WATER BREAK!

It's important to stay hydrated.

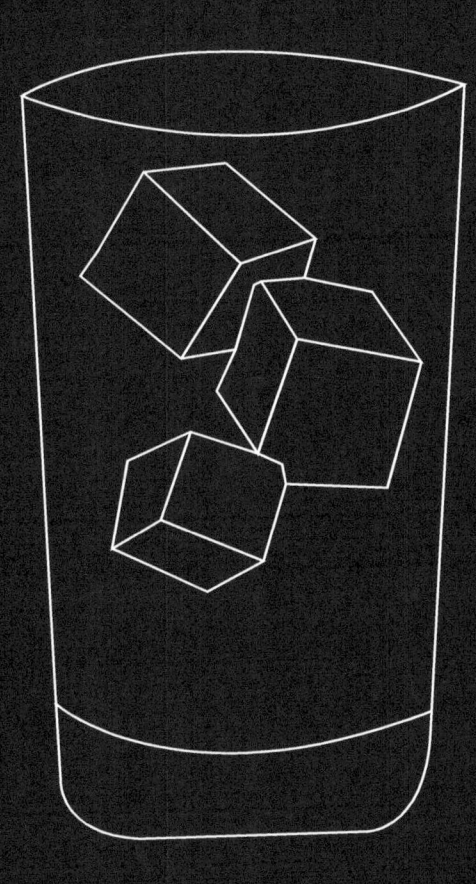

CONNECT THE CONSTELLATION

Knowing what you're looking at is
important for galactic navigation.

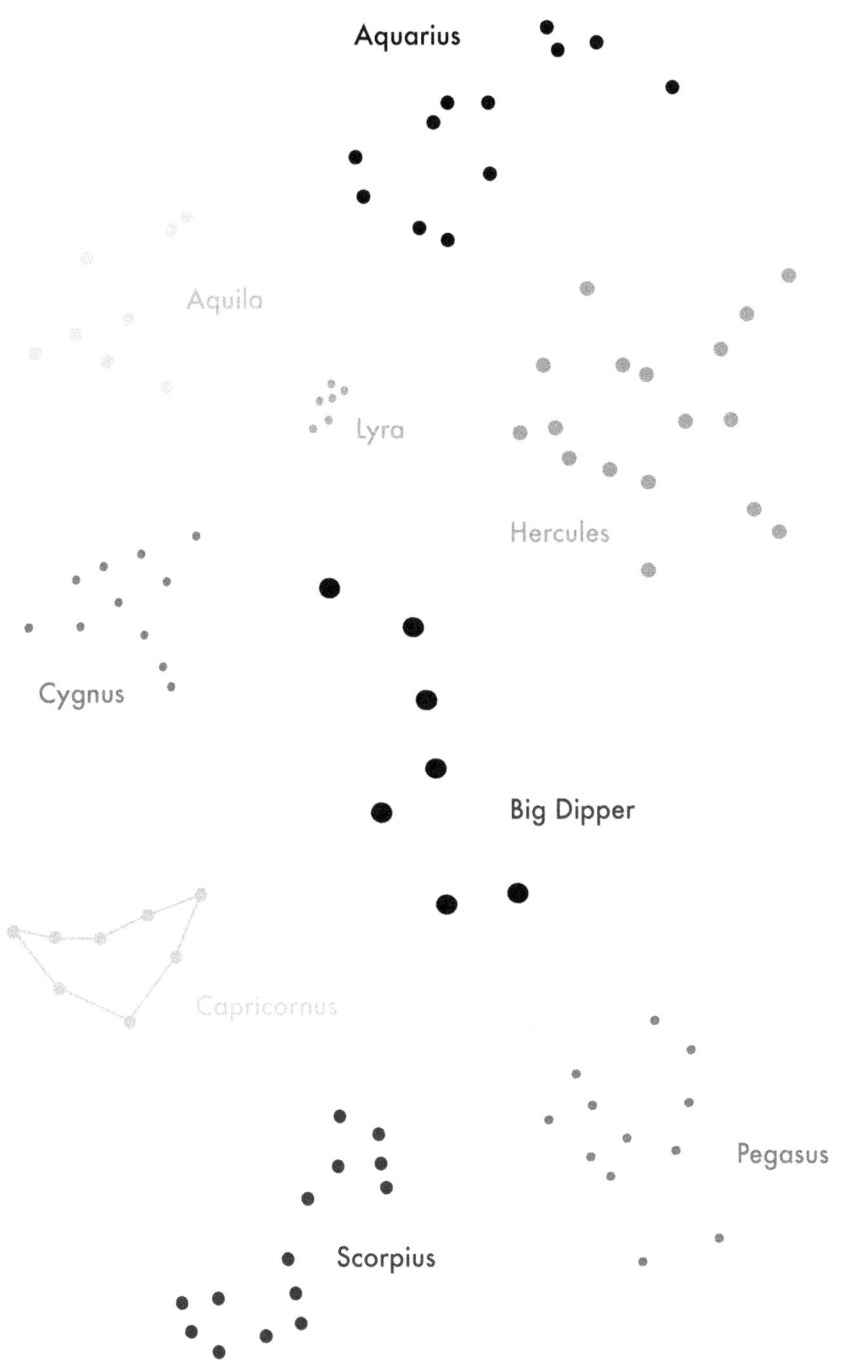

Aquarius

Aquila

Lyra

Hercules

Cygnus

Big Dipper

Capricornus

Pegasus

Scorpius

HERE MEN FROM THE PLANET EARTH FIRST SET FOOT UPON THE MOON JULY 1969, A.D. WE CAME IN PEACE FOR ALL MANKIND

Neil Armstrong and Buzz Aldrin didn't just leave footprints on the moon, they also left a flag and a plaque with a peaceful message etched on the front for any life forms that inhabit the moon.

WHAT WILL YOUR PLAQUE SAY?

GLUE A PIECE OF EARTH HERE

You won't be seeing many trees or flowers when you travel to outer space, so you should bring a little bit of Earth with you so you don't get homesick.

DIRT

AND HERE

 LEAF

STICK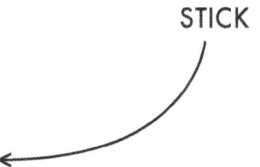

HUNGRY?

Tang is a fruit-flavored drink made from mixing powder and water together. This colorful drink was first used on John Glenn's Mercury flight and has continued to be a favorite on board manned space flights.

Make your own:

- Tang powder
- 1 cup of water

Mix the drink in a ziplock bag and sip with a straw to really feel like an astronaut.

How did it taste?

_____

_____

_____

_____

_____

_____

FILL THIS PAGE WITH COMETS

A comet is a small icy ball that
grows a tail when it gets too hot.

HAVE YOU EVER SEEN A COMET?

STOP AND TOUCH YOUR TOES

It's important to be flexible so you don't
get cramps while floating in space.

I TOUCHED MY TOES
YES/ALMOST/NO

WATER BREAK!

It's important to stay hydrated.

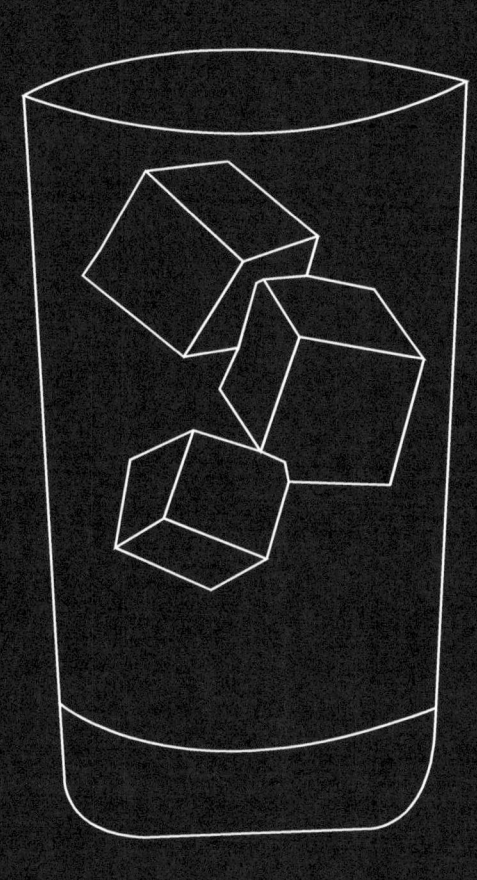

DON'T FORGET TO WRITE

Cut out this postcard while you're traveling through space and send it back on a cargo ship to Earth for your family and friends.

GREETINGS FROM OUTER SPACE!

DON'T KNOW WHAT TO WRITE?

Write about all the stars you've counted or which planets you've been to so far. You can also draw a picture of you with your friends and family in outer space together!

SO YOU RUN INTO AN ALIEN

What do you talk about with your new friend?

_____

_____

_____

_____

_____

_____

FIVE OR SIX EYES?

What does your new friend look like?

HUNGRY?

Chocolate pudding is a great food for astronauts to eat during their voyage because it's soft and easy to consume through a straw. Sometimes caffeine is added for an extra kick!

Make your own:

- 2 tbsp instant pudding mix
- 2 tbsp powdered milk
- 1/2 cup water

Mix the pudding in a ziplock bag and consume with a straw to really feel like an astronaut.

How did it taste?

_____

_____

_____

_____

_____

IF YOU GET BORED

The moon is about 250,000 miles away so here's
something to keep you busy on your way there.

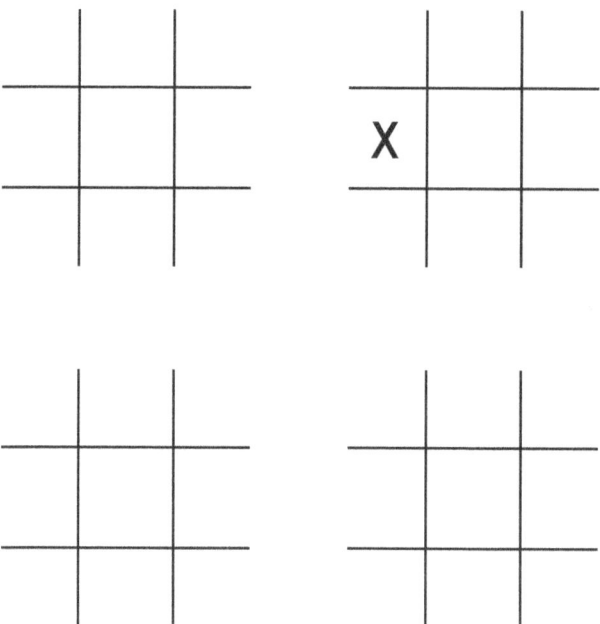

```
X W S E Q H I I T K T B G R C
G Q A R C Z O U N I V E R S E
G R L T U C A H R J B N S P X
Y V A F E N T L P N E C H P U
Z X P V O R U I S E A R T H E
E I A R I N F F U H V M J N A
L Q T L E T M T N H N K G O T
Z S B I A I Y O R G H P M O E
A F L P S G Z F O T N A L M N
G A J S U Y L F C F R A I H A
Q D I J R A T S I S D G T X L
G O C V F W Z F R H G H E L P
N E L T T U H S P O B L M Q O
D Y V K K C U P A M Y L H S V
C G D T V T T A C E F L O F A
```

ALIEN HOME SHUTTLE
ASTRONAUT LIFTOFF STAR
CAPRICORNUS MARS TANG
EARTH MISSION UNIVERSE
GALAXY MOON WATER
GRAVITY PLANET

IF TIMES GET TOUGH

Morale can get low when you are away from home for months at a time. Humor is a great way to lighten up the mood and make operations run smoothly again.

Q: What is a spaceman's favorite chocolate?
A: A Mars bar!

Q: What kind of music do planets sing?
A: Neptunes!

Q: What do planets like to read?
A: Comet books!

Q: What is an astronaut's favorite key on the keyboard?
A: The space bar!

Q: How do you know when the moon has enough to eat?
A: When it's full!

Q: _____
A: _____

Q: _____
A: _____

CHECKLIST

Astronauts follow their checklists very closely before liftoff, during their mission and before landing.

☐ SPACE SUIT FITS PROPERLY

☐ FOOD AND BELONGINGS ARE LOADED

☐ MUSCLES ARE FIT FOR THE RIDE

☐ TENSION RELIEVERS ARE IN MIND

☐ JOURNEY IS MAPPED OUT

☐ NIGHT SKY ACTIVITY IS NOTED

☐ REVIEW KNOWLEDGE OF CONSTELLATIONS

☐ BUCKLE UP FOR A FAST RIDE

Cut out your patch and wear it proudly.

YOU'RE READY FOR LIFTOFF

WWW.

To learn more before liftoff grab a parent
and check out the following websites.

nasa.gov

space.com

challenger.org

science.nationalgeographic.com

pbs.org

popsci.com

NOTES

_____

_____

_____

_____

_____

_____

_____

_____

_____

NOTES

_____

_____

_____

_____

_____

_____

_____

_____

_____

NOTES

_____

_____

_____

_____

_____

_____

_____

_____

_____

NOTES

WAIT! Before you go you should know that outer space
is only a little over 60 miles away. Not so far after all.

60

www.ingramcontent.com/pod-product-compliance
Lightning Source LLC
Chambersburg PA
CBHW051820170526
45167CB00005B/2089